info@ashleysnotebooks.com
www.ashleysnotebooks.com